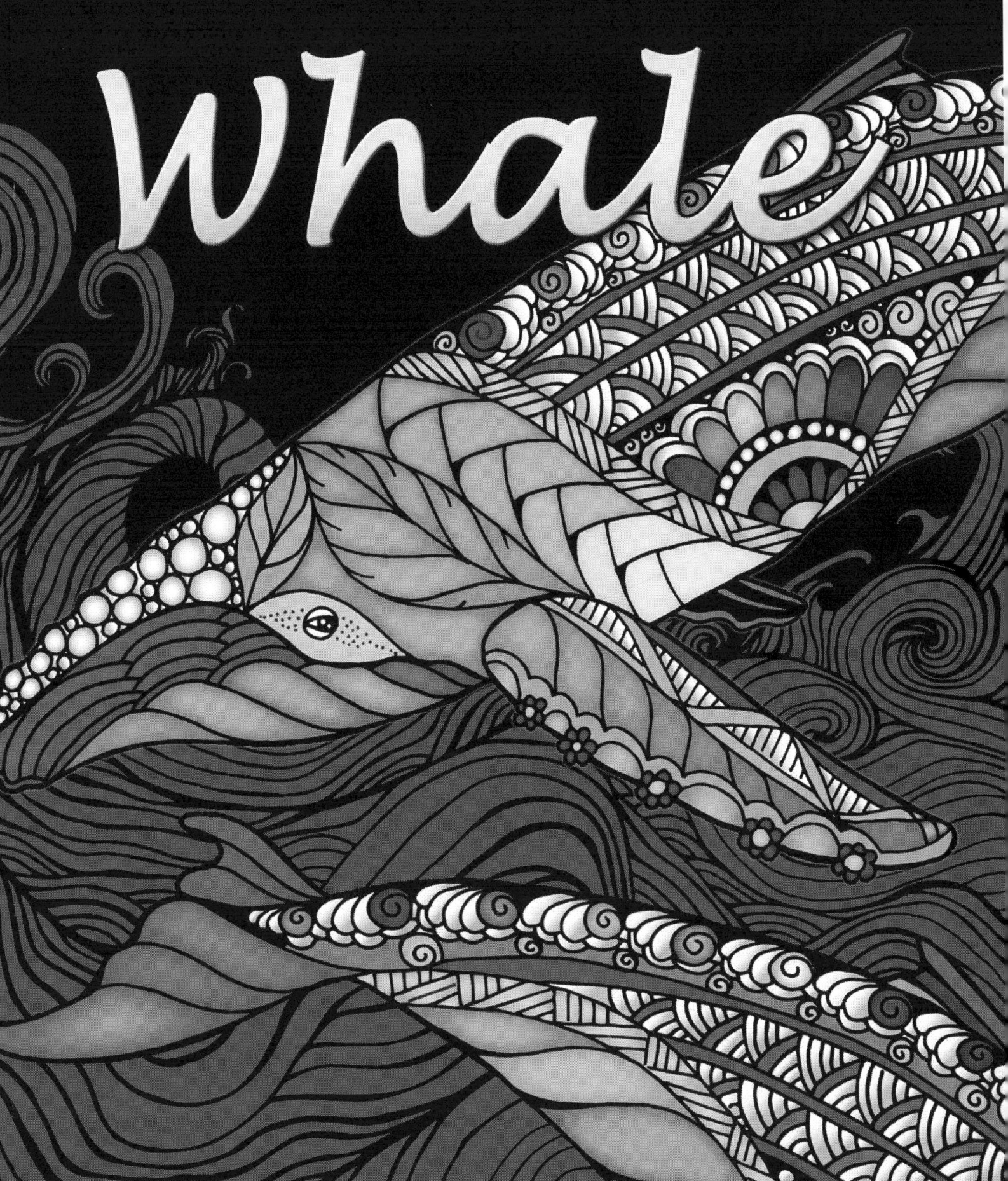

Whale

COLORING BOOK FOR ADULT

PUBLISHED IN 2018 BY
KODOMO PUBLISHING

PRINTED IN THE UNITED STATES OF AMERICA

COLOR TEST PAGE

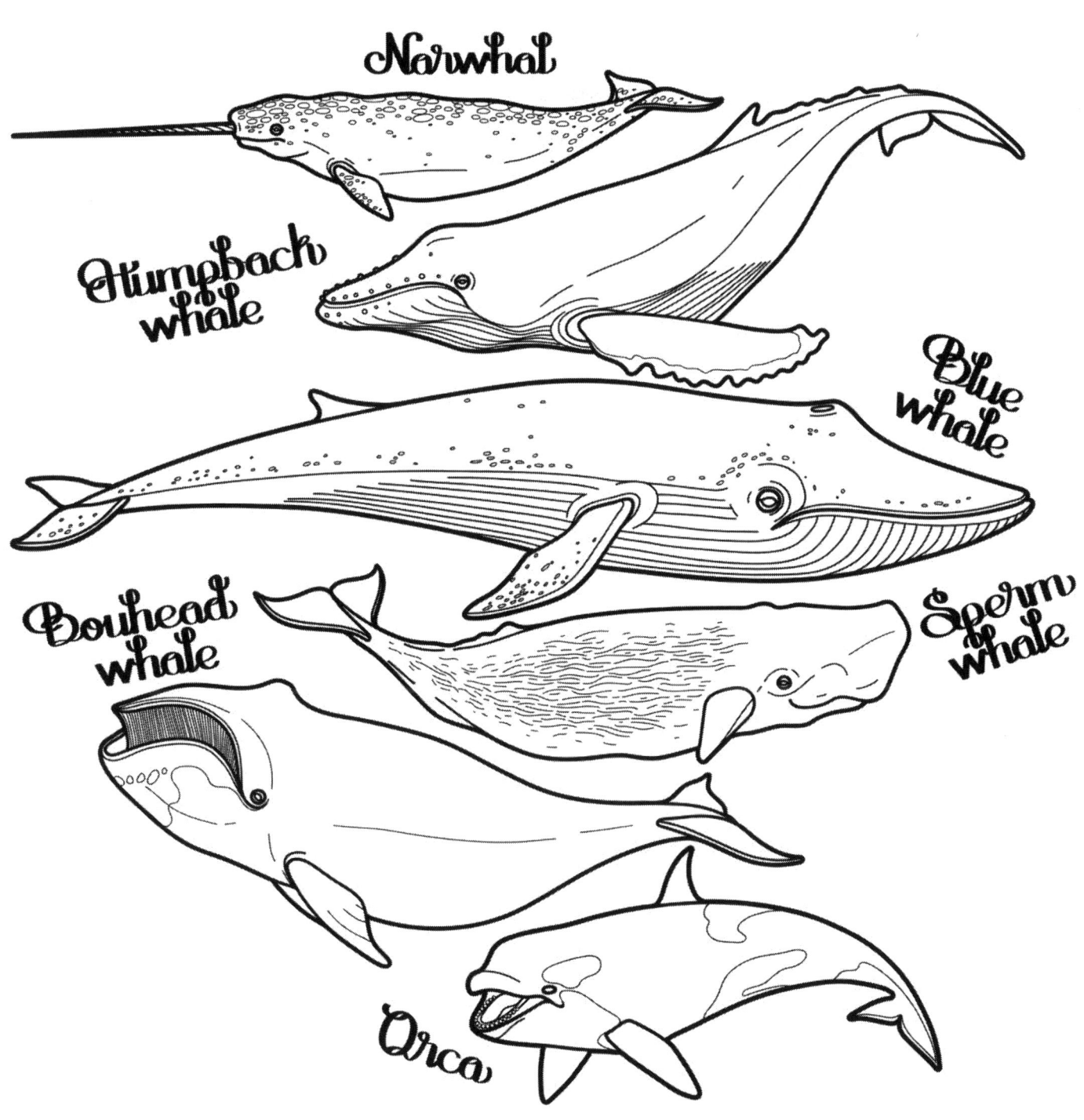

Narwhal

Humpback whale

Blue whale

Bowhead whale

Sperm whale

Orca

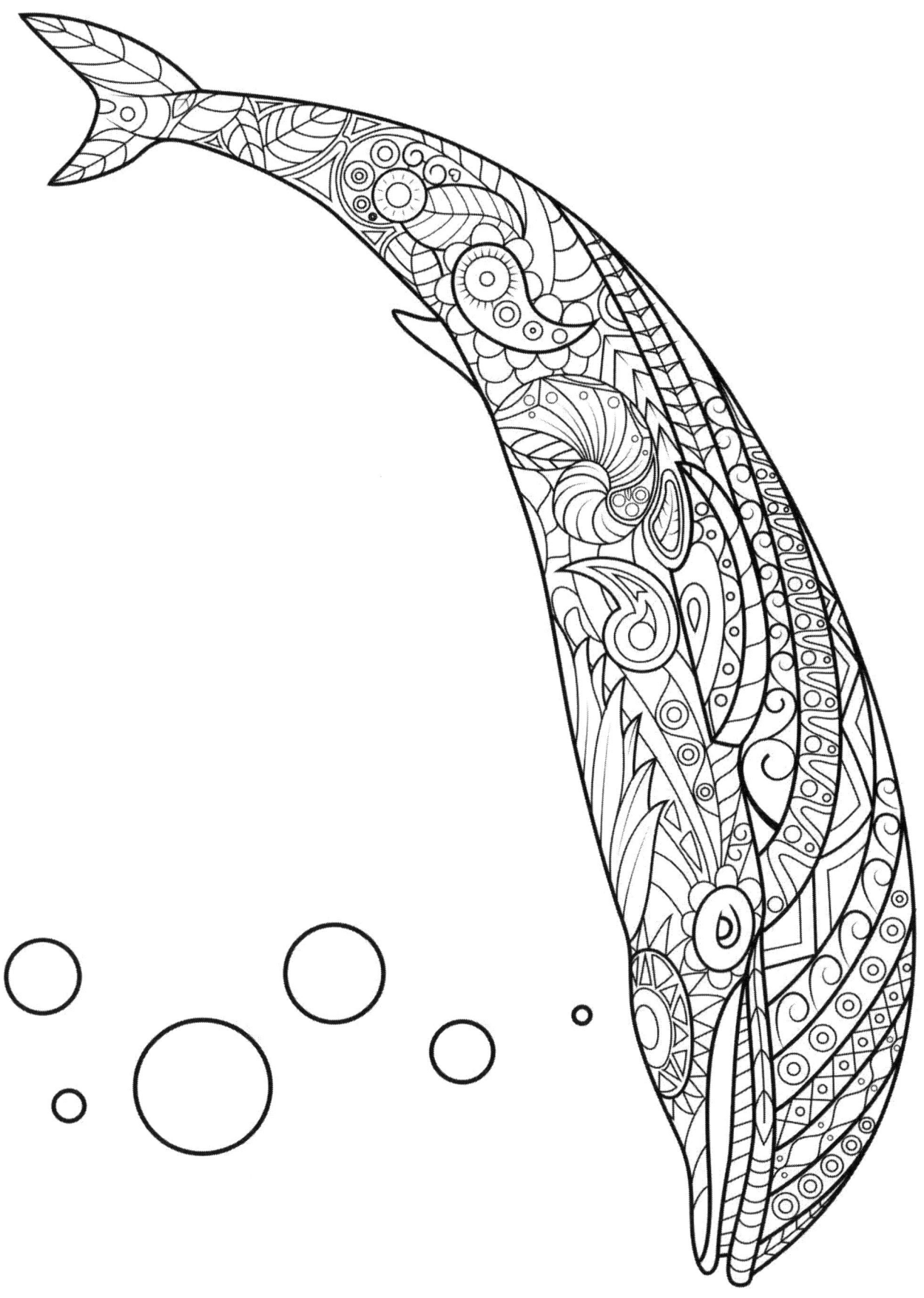

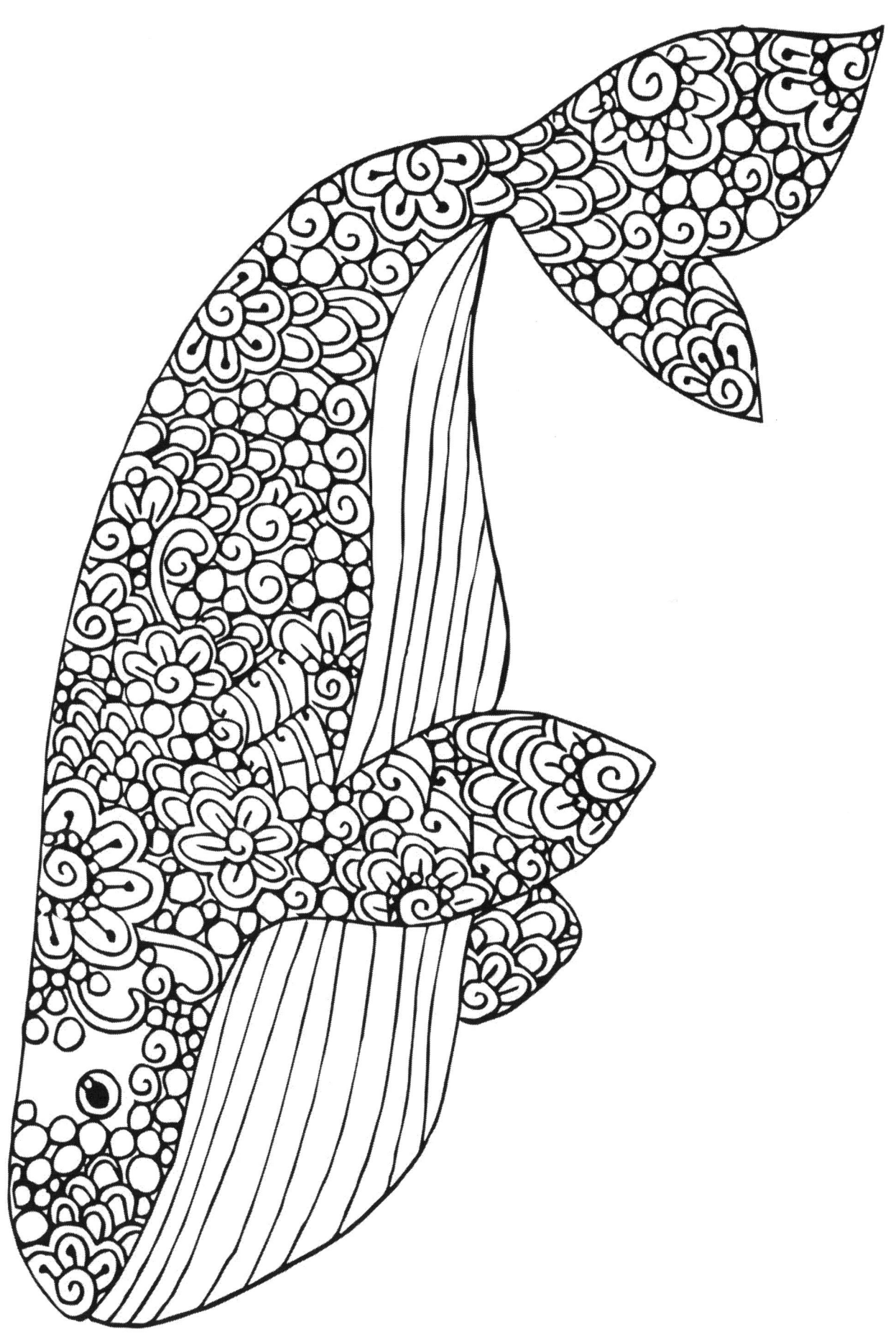

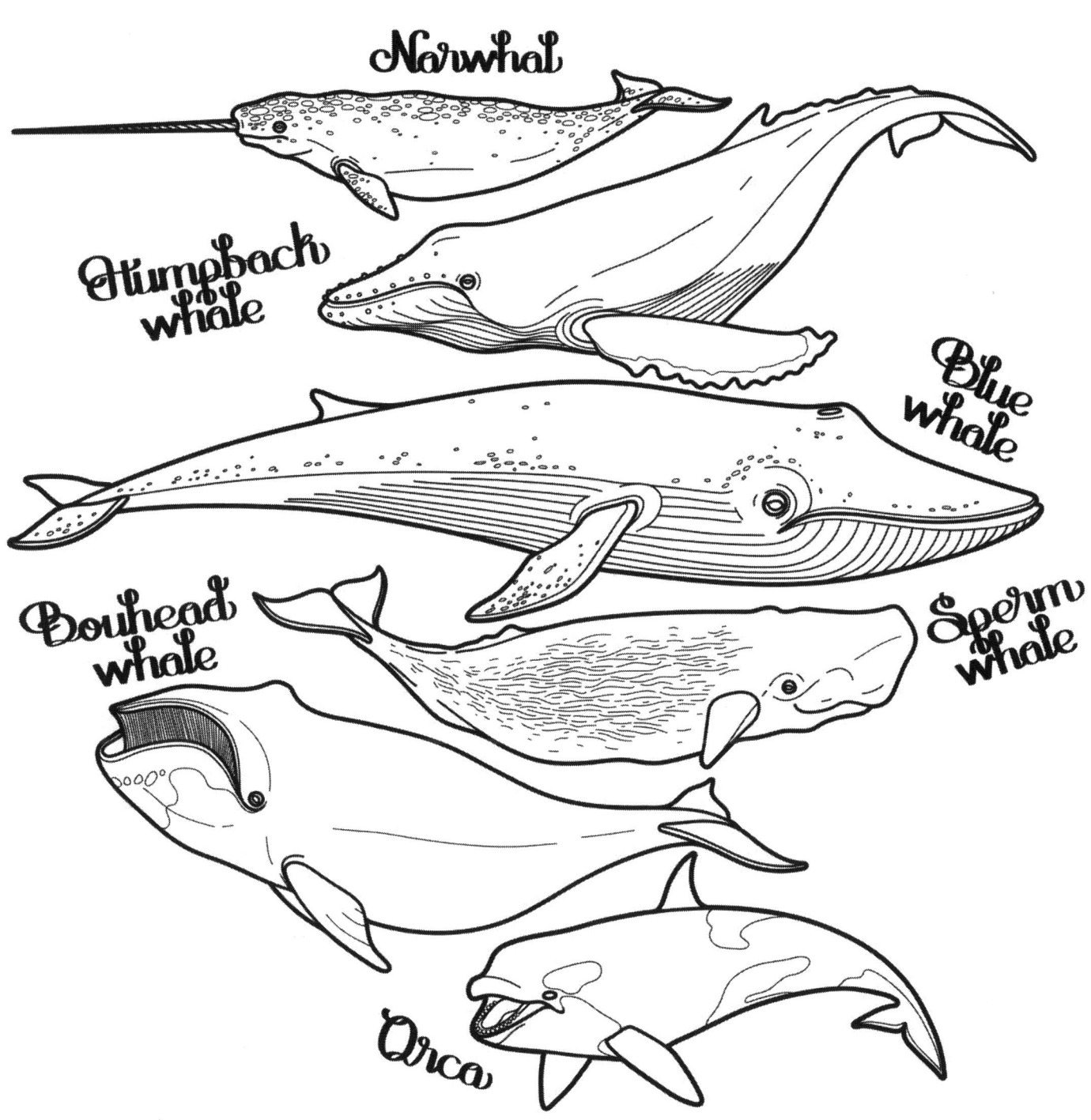

Narwhal

Humpback whale

Blue whale

Bowhead whale

Sperm whale

Orca

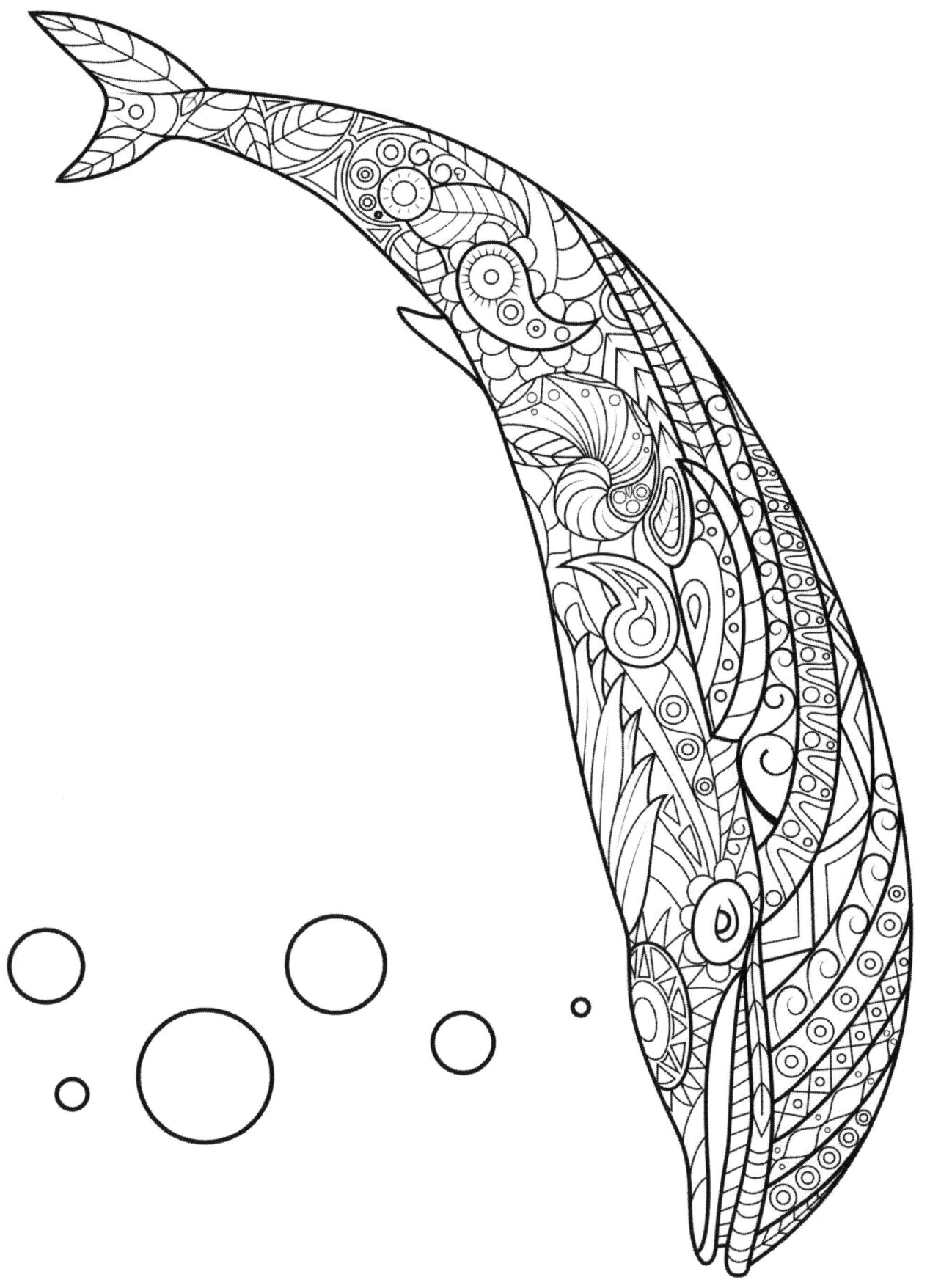

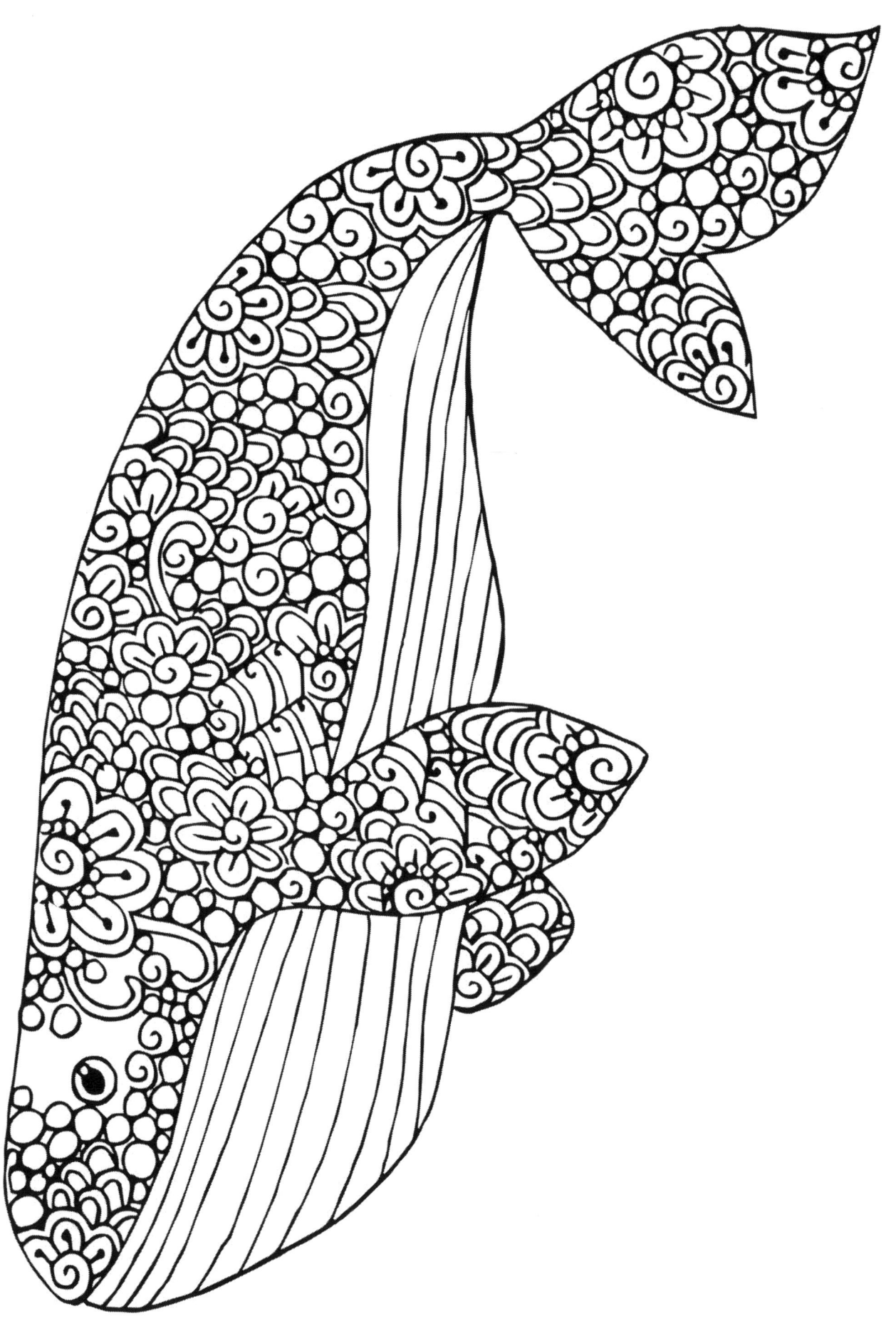

Printed in Great Britain
by Amazon